MENU MATH

presents

The Old-Fashioned Ice Cream Parlor

Book 1

A Remedia Publication

Created by
Barbara Johnson
Kitty Scharf

Illustrations by
Edwin Vandermulen

FOR INFORMATION ABOUT OUR SOFTWARE OR CATALOG
CALL 1-800-826-4740 OR 602-994-5909

REMEDIA PUBLICATIONS 10135 E. VIA LINDA #D124 SCOTTSDALE, AZ 85258
602-661-9900 TOLL FREE 1-800-826-4740 FAX 602-661-9901

To the teacher: For additional practice, blank checks are provided so students may create their own using the menu and the tax table on page 36.

Ice Cream Parlor		
	Subtotal	
	Tax	
	Total	

Ice Cream Parlor		
	Subtotal	
	Tax	
	Total	

Ice Cream Parlor		
	Subtotal	
	Tax	
	Total	

Ice Cream Parlor		
	Subtotal	
	Tax	
	Total	

Name ______________________________

Introducing decimal notation for money

EXAMPLE: 64¢ can be written as...

$ .64 ← 4 pennies
- 6 dimes
- decimal point
- dollar sign

Write the amounts below using the dollar sign ($) and the decimal point (.).

1. three cents ________
2. seven cents ________
3. twelve cents ________
4. seventeen cents ________
5. thirty-five cents ________
6. sixty cents ________
7. ninety-nine cents ________
8. one dollar and fourteen cents ________
9. twelve dollars and twenty-two cents ________
10. thirty dollars and twenty-nine cents ________
11. forty-five dollars and sixteen cents ________
12. fifty-two dollars and two cents ________
13. sixty dollars and forty-five cents ________
14. seventy-five dollars and twenty cents ________

Name ______________________

Adding two 2-digit numbers, no regrouping

EXAMPLE:

$$\begin{array}{r} \$\ .17 \\ +\ .01 \\ \hline \$\ .18 \end{array}$$

$ .11 + .05	$ .16 + .01	$ .31 + .06	$ .14 + .10
$ .15 + .30	$ .12 + .24	$ .44 + .15	$ .32 + .11
$ .10 + .25	$ .12 + .26	$ .44 + .22	$ .60 + .20
$ .40 + .20	$ .25 + .14	$ .69 + .10	$ .44 + .31
$ .26 + .23	$ .41 + .24	$ .82 + .15	$ .47 + .21
$ .20 + .15	$ .21 + .10	$ .92 + .02	$ .86 + .12

Name ______________________________

Adding two
2-digit numbers,
no regrouping

$ .59 + .20	$ .61 + .23	$ .42 + .13	$ .56 + .23
$ .68 + .10	$ .71 + .25	$ .62 + .16	$ .69 + .10
$ .43 + .32	$ .76 + .12	$ .73 + .11	$ .70 + .12
$ .81 + .14	$ .83 + .14	$ .84 + .12	$ .80 + .14
$ .46 + .20	$ .51 + .06	$ .66 + .21	$ .32 + .12
$ .91 + .03	$ .86 + .10	$ .80 + .06	$ .43 + .12

Name ____________________________

Adding three 2-digit numbers, no regrouping

Example:

$ .11
.21
+ .10
$ .42

$ .21 .42 + .14	$ .27 .32 + .40	$ .20 .18 + .31	$ .30 .32 + .02
$ .10 .22 + .25	$ .24 .13 + .61	$ .63 .14 + .12	$ .61 .10 + .10
$ .20 .41 + .14	$ .25 .42 + .12	$ .31 .33 + .25	$ .52 .13 + .11
$ .10 .21 + .25	$ .16 .12 + .11	$ .58 .11 + .20	$ .47 .10 + .12
$ .20 .12 + .32	$ .53 .21 + .15	$ .43 .12 + .31	$ .33 .12 + .40
$ .10 .25 + .40	$ .17 .32 + .50	$ .23 .53 + .21	$ .42 .13 + .12

Name ______________________________

**Adding three
2-digit numbers,
no regrouping**

$ .13 .25 + .41	$ .15 .20 + .42	$ .31 .25 + .13	$ .22 .32 + .25
$ .26 .20 + .11	$ .26 .12 + .10	$ .35 .31 + .10	$ .42 .10 + .15
$ .12 .21 + .23	$ .23 .34 + .11	$ .52 .10 + .22	$ .55 .14 + .10
$ .23 .42 + .14	$ .18 .20 + .11	$ .62 .12 + .14	$ .15 .22 + .60
$ .16 .21 + .22	$ .13 .15 + .11	$ .36 .11 + .12	$ .50 .16 + .12
$ .20 .17 + .31	$ .19 .20 + .30	$ .66 .10 + .13	$ .82 .10 + .02

Name ______________________________

Adding two 3-digit numbers, no regrouping

EXAMPLE:

$$\begin{array}{r} \$\ 1.22 \\ +\ 1.10 \\ \hline \$\ 2.32 \end{array}$$

\$ 1.22 + 1.31	\$ 1.23 + 1.22	\$ 2.15 + 3.04	\$ 3.30 + 1.20
\$ 3.16 + 2.01	\$ 1.26 + 2.13	\$ 2.03 + 4.15	\$ 5.21 + 2.06
\$ 5.52 + 3.21	\$ 2.14 + 2.13	\$ 4.05 + 2.01	\$ 4.02 + 4.03
\$ 4.31 + 2.30	\$ 6.60 + 2.21	\$ 6.52 + 2.20	\$ 3.52 + 5.06
\$ 6.50 + 2.23	\$ 3.09 + 2.00	\$ 7.01 + 2.80	\$ 6.40 + 1.30
\$ 5.14 + 2.01	\$ 1.20 + 1.35	\$ 7.03 + 2.05	\$ 8.10 + 1.60

Name ________________________________

Adding three 3-digit numbers, no regrouping

EXAMPLE:

$ 2.15
2.21
+ 1.53
$ 5.89

$ 2.13	$ 3.21	$ 3.06	$ 4.41
2.22	1.22	2.00	2.10
+ 1.50	+ 2.50	+ 3.60	+ 3.20
$ 4.23	$ 5.24	$ 2.51	$ 5.40
2.22	3.11	2.31	2.21
+ 1.20	+ 1.62	+ 3.00	+ 2.32
$ 6.12	$ 7.14	$ 4.15	$ 7.12
1.30	1.61	1.20	1.32
+ 2.51	+ 1.11	+ 1.10	+ 1.12
$ 4.12	$ 6.13	$ 4.24	$ 3.20
1.22	1.31	2.22	2.14
+ 3.12	+ 2.42	+ 1.23	+ 1.14
$ 2.04	$ 3.41	$ 1.51	$ 2.14
1.21	2.02	3.26	4.50
+ 3.40	+ 2.01	+ 2.11	+ 3.23
$ 4.52	$ 5.01	$ 1.26	$ 1.66
2.33	1.24	2.21	3.10
+ 1.02	+ 3.71	+ 4.52	+ 5.21

Name ______________________________

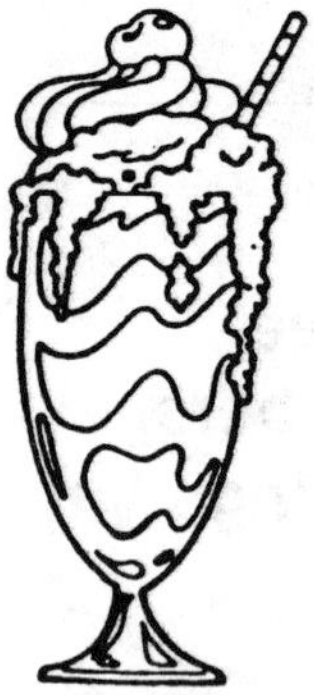

Adding three
3-digit numbers,
no regrouping

\$ 2.50 1.32 + 6.12	\$ 1.12 1.61 + 7.13	\$ 4.30 2.10 + 3.21	\$ 1.10 1.51 + 7.12
\$ 3.06 1.22 + 4.11	\$ 2.42 1.31 + 6.12	\$ 1.66 3.10 + 5.20	\$ 1.52 2.22 + 1.23
\$ 3.36 1.23 + 2.00	\$ 2.01 2.03 + 3.41	\$ 1.41 1.21 + 3.11	\$ 3.16 1.12 + 3.10
\$ 1.00 2.35 + 4.50	\$ 3.71 1.24 + 5.02	\$ 7.10 1.25 + 1.30	\$ 5.41 2.16 + 1.21
\$ 3.61 2.02 + 3.04	\$ 3.20 2.11 + 4.51	\$ 6.12 1.02 + 2.15	\$ 4.16 1.02 + 3.00
\$ 3.01 2.34 + 2.51	\$ 2.35 2.20 + 5.40	\$ 6.00 1.15 + 2.04	\$ 4.00 2.02 + 3.17

Name ______________________________

Adding two 2-digit numbers, one-step regrouping

EXAMPLE:

1
$ .38
+ .13
$ .51

$ .36 + .14	$ .52 + .18	$ .62 + .28	$ .29 + .34
$ .19 + .21	$ .34 + .26	$ .63 + .19	$ .55 + .35
$ .59 + .11	$ .48 + .13	$ .58 + .29	$ .74 + .16
$ .29 + .15	$ .48 + .15	$ .86 + .09	$ .44 + .08
$ .33 + .27	$ .36 + .18	$ .27 + .13	$ .35 + .16
$ .19 + .25	$ .59 + .29	$ .25 + .18	$ .26 + .36

Name ______________________

Adding two
2-digit numbers,
one-step regrouping

$ .13 + .17	$ .12 + .18	$ .19 + .15	$ .27 + .16
$ .12 + .19	$ .15 + .25	$ .34 + .29	$ .76 + .14
$ .17 + .24	$ .13 + .19	$ .45 + .36	$ .28 + .32
$ .13 + .18	$ .16 + .14	$ .46 + .46	$ .39 + .11
$ .22 + .18	$ .24 + .19	$ .52 + .38	$ .66 + .24
$ .25 + .35	$ .17 + .17	$ .36 + .15	$ .29 + .12

Name ______________________________

Adding two 2-digit numbers, two-step regrouping

EXAMPLE:

1
$ .96
+ .94
$ 1.90

$ .28 + .82	$ .68 + .73	$ .84 + .76	$ .69 + .35
$ .29 + .71	$ .99 + .86	$ .48 + .53	$ .49 + .64
$ .38 + .74	$ .46 + .84	$ .75 + .95	$ .95 + .27
$ .37 + .73	$ .63 + .89	$ .94 + .88	$ .91 + .99
$ .55 + .65	$ .68 + .52	$ .59 + .54	$ .97 + .83
$ .56 + .48	$ .29 + .84	$ .86 + .58	$ .93 + .57

Name ______________________________

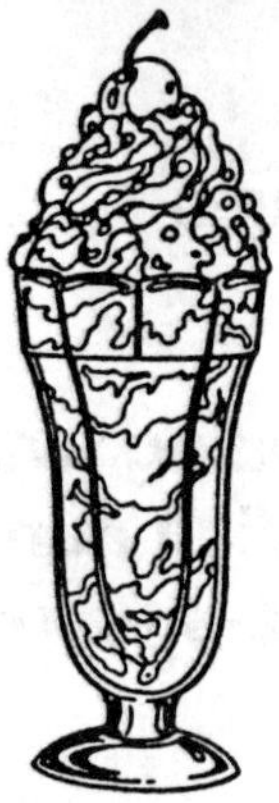

Adding two 2-digit numbers, two-step regrouping

$ .84 + .98	$ .55 + .57	$ .61 + .99	$ .69 + .88
$ .78 + .83	$ .84 + .89	$ .58 + .57	$ .87 + .83
$ .94 + .78	$ .68 + .89	$ .88 + .85	$ .18 + .94
$ .38 + .83	$ .65 + .95	$ .63 + .48	$ .87 + .46
$ .48 + .89	$ .29 + .79	$ .75 + .95	$ .78 + .38
$ .88 + .36	$ .45 + .98	$ .77 + .44	$ .67 + .43

Name ________________________________

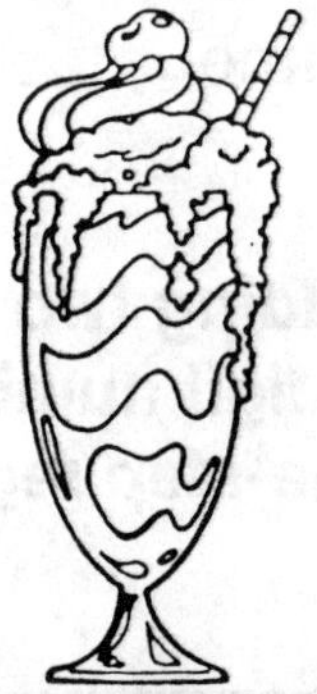

Adding two
2-digit numbers,
two-step regrouping

$ .96 + .87	$ .94 + .67	$ .66 + .46	$ .29 + .93
$ .74 + .36	$ .72 + .48	$ .27 + .76	$ .32 + .69
$ .77 + .67	$ .32 + .79	$ .85 + .39	$ .27 + .85
$ .43 + .78	$ .92 + .39	$ .91 + .39	$ .53 + .78
$ .48 + .84	$ .76 + .66	$ .66 + .66	$ .88 + .88
$ .49 + .86	$ .88 + .57	$ .99 + .99	$ .75 + .85

Name ______________________________

Adding two 3-digit numbers, one-step regrouping

EXAMPLE:

1
\$ 1.16
+ 1.19
\$ 2.35

\$ 1.12 + 1.19	\$ 1.24 + 1.36	\$ 4.13 + 4.17	\$ 6.49 + 2.21
\$ 3.15 + 2.15	\$ 4.35 + 2.35	\$ 3.58 + 5.04	\$ 7.72 + 2.19
\$ 5.53 + 3.27	\$ 6.66 + 3.17	\$ 1.14 + 1.19	\$ 1.37 + 1.24
\$ 6.69 + 3.21	\$ 5.34 + 4.27	\$ 2.18 + 3.02	\$ 2.33 + 4.37
\$ 1.21 + 1.39	\$ 2.14 + 2.26	\$ 5.36 + 2.24	\$ 6.29 + 2.11
\$ 2.17 + 4.16	\$ 5.26 + 3.04	\$ 5.46 + 2.14	\$ 3.59 + 1.11

Name ______________________________

Adding two
3-digit numbers,
one-step regrouping

$ 4.36 + 2.27	$ 2.67 + 4.19	$ 6.18 + 2.09	$ 7.72 + 1.18
$ 5.21 + 4.59	$ 6.18 + 2.19	$ 7.14 + 1.18	$ 7.19 + 2.19
$ 8.02 + 1.08	$ 6.09 + 3.19	$ 7.61 + 1.09	$ 5.49 + 2.18
$ 8.14 + 1.56	$ 1.57 + 8.14	$ 6.14 + 2.19	$ 8.18 + 1.23
$ 6.15 + 2.16	$ 3.08 + 6.09	$ 8.28 + 1.26	$ 8.15 + 1.15
$ 4.18 + 5.78	$ 3.13 + 4.19	$ 8.16 + 1.19	$ 7.16 + 2.24

Name ______________________________

Adding two 3-digit numbers, two-step regrouping

EXAMPLE:

```
   1 1
$ 1.67
+ 2.33
------
$ 4.00
```

$ 1.68 + 2.34	$ 1.36 + 4.94	$ 5.36 + 1.87	$ 6.29 + 1.89
$ 1.18 + 2.99	$ 1.56 + 3.98	$ 1.95 + 2.85	$ 2.39 + 1.67
$ 1.89 + 2.56	$ 2.87 + 2.23	$ 6.15 + 2.89	$ 3.33 + 2.77
$ 2.65 + 4.95	$ 2.79 + 2.62	$ 4.93 + 4.08	$ 6.48 + 2.65
$ 4.53 + 2.98	$ 6.98 + 2.18	$ 2.77 + 2.33	$ 1.56 + 4.94
$ 4.38 + 2.65	$ 6.64 + 2.37	$ 4.39 + 2.66	$ 6.49 + 1.78

Name ______________________________

Adding two
3-digit numbers,
two-step regrouping

$ 6.18 + 2.97	$ 2.93 + 2.07	$ 7.52 + 1.98	$ 1.77 + 5.54
$ 3.99 + 2.09	$ 5.99 + 2.08	$ 3.55 + 1.66	$ 1.89 + 7.39
$ 2.97 + 1.99	$ 1.98 + 5.69	$ 3.49 + 1.78	$ 5.85 + 1.67
$ 7.84 + 1.99	$ 7.14 + 1.99	$ 1.63 + 7.59	$ 6.59 + 1.95
$ 2.99 + 3.76	$ 6.52 + 2.89	$ 1.75 + 3.46	$ 3.58 + 3.45
$ 5.95 + 3.95	$ 1.67 + 6.98	$ 6.38 + 2.67	$ 7.55 + 1.45

Name ______________________________

Adding two 3-digit numbers, two or more regrouping steps

EXAMPLE:

 1 1
 $ 7.36
+ 6.84
$14.20

$ 8.58 + 6.84	$ 4.69 + 8.96	$ 4.09 + 9.99	$ 6.78 + 8.76
$ 7.49 + 6.75	$ 4.98 + 6.46	$ 4.12 + 9.98	$ 7.29 + 8.99
$ 5.98 + 8.27	$ 6.09 + 9.97	$ 6.37 + 8.68	$ 4.03 + 9.97
$ 8.23 + 9.79	$ 6.36 + 4.67	$ 8.88 + 9.97	$ 6.64 + 8.99
$ 1.91 + 9.09	$ 4.89 + 9.48	$ 5.55 + 5.46	$ 6.29 + 7.71
$ 3.86 + 9.79	$ 6.86 + 6.86	$ 6.59 + 6.75	$ 4.46 + 7.65

Name ______________________________

Adding two 3-digit numbers, two or more regrouping steps

$ 8.38 + 2.83	$ 6.49 + 6.96	$ 2.04 + 9.97	$ 6.27 + 8.95
$ 8.29 + 9.93	$ 6.29 + 5.85	$ 7.35 + 5.85	$ 6.67 + 7.67
$ 8.67 + 7.77	$ 8.14 + 4.86	$ 8.04 + 9.97	$ 3.99 + 8.34
$ 8.39 + 5.69	$ 7.87 + 9.95	$ 6.96 + 5.24	$ 3.97 + 7.98
$ 5.66 + 8.88	$ 1.11 + 9.99	$ 6.43 + 4.57	$ 5.96 + 4.35
$ 9.69 + 6.96	$ 8.33 + 5.78	$ 3.67 + 8.54	$ 6.66 + 7.77

Name ______________________________

Adding two 4-digit numbers, two or more regrouping steps

EXAMPLE:

11 1
$ 42.37
+ 19.83
$ 62.20

$ 34.76 + 15.77	$ 40.67 + 29.83	$ 58.65 + 49.45	$ 41.68 + 29.37
$ 50.57 + 34.65	$ 30.19 + 25.83	$ 24.53 + 17.87	$ 33.75 + 16.85
$ 70.93 + 19.18	$ 85.67 + 12.85	$ 13.59 + 11.67	$ 72.41 + 56.89
$ 82.16 + 48.94	$ 20.08 + 19.98	$ 54.84 + 26.76	$ 32.84 + 15.29
$ 77.67 + 15.68	$ 28.39 + 15.86	$ 34.68 + 18.82	$ 68.62 + 19.78
$ 44.72 + 26.48	$ 36.66 + 29.84	$ 36.36 + 28.69	$ 51.81 + 29.89

Name ______________________________

Adding two 4-digit numbers, two or more regrouping steps

$ 36.85 + 18.78	$ 24.66 + 28.48	$ 61.68 + 18.79	$ 52.56 + 34.56
$ 78.79 + 29.36	$ 76.83 + 19.97	$ 68.78 + 37.67	$ 20.16 + 19.96
$ 86.15 + 54.96	$ 69.15 + 48.95	$ 76.29 + 36.87	$ 76.68 + 56.87
$ 66.58 + 56.67	$ 82.82 + 78.68	$ 87.86 + 29.39	$ 62.15 + 48.96
$ 88.68 + 52.39	$ 72.76 + 59.36	$ 98.16 + 79.95	$ 68.98 + 39.85
$ 82.15 + 68.96	$ 40.98 + 29.76	$ 82.32 + 68.78	$ 63.16 + 54.85

Name ______________________________

Adding three 3-digit numbers, two or more regrouping steps

EXAMPLE:

```
   1 1
 $ 6.05
   3.05
 + 1.98
 ------
$ 11.08
```

\$ 3.57 2.69 + 4.53	\$ 2.15 6.86 + 1.63	\$ 4.14 2.57 + 3.43	\$ 2.89 5.09 + 3.12
\$ 3.52 3.64 + 3.04	\$ 5.40 2.28 + 3.92	\$ 3.19 2.98 + 4.64	\$ 2.15 8.50 + 2.46
\$ 1.56 3.85 + 6.01	\$ 2.61 2.89 + 5.10	\$ 8.91 3.47 + 2.05	\$ 9.10 2.98 + 1.57
\$ 6.04 1.98 + 3.05	\$ 8.64 2.43 + 1.34	\$ 5.07 3.84 + 2.23	\$ 4.18 5.10 + 2.98
\$ 3.98 2.01 + 4.12	\$ 3.06 2.98 + 4.15	\$ 6.14 3.59 + 2.46	\$ 3.23 4.25 + 5.53
\$ 3.06 2.97 + 4.18	\$ 6.09 2.15 + 2.84	\$ 8.14 3.51 + 4.86	\$ 1.57 3.86 + 5.50

Name ______________________________

Adding three
3-digit numbers,
two or more
regrouping steps

$ 2.56 3.94 + 4.02	$ 4.38 2.50 + 5.42	$ 6.93 2.08 + 2.25	$ 1.69 8.50 + 2.43
$ 9.10 2.72 + 1.58	$ 3.69 2.50 + 5.02	$ 7.91 2.98 + 3.11	$ 8.50 4.99 + 3.24
$ 6.15 1.44 + 2.62	$ 4.14 6.85 + 1.24	$ 8.04 2.43 + 2.66	$ 2.57 3.53 + 5.50
$ 8.52 3.68 + 1.50	$ 1.99 2.01 + 7.23	$ 2.16 4.77 + 3.27	$ 9.28 1.57 + 2.33
$ 3.06 5.94 + 3.21	$ 1.57 1.51 + 8.02	$ 2.54 5.66 + 3.02	$ 2.37 4.52 + 3.42
$ 2.89 2.64 + 6.01	$ 4.48 2.96 + 5.10	$ 1.85 5.61 + 4.57	$ 6.68 1.46 + 3.25

Name ______________________________

EXAMPLE:

Adding three 4-digit numbers, two or more regrouping steps

```
    21 1
  $ 19.23
    28.50
  + 33.47
  -------
  $ 81.20
```

\$ 33.64 30.06 + 21.35	\$ 14.27 54.68 + 34.21	\$ 79.31 13.79 + 19.25	\$ 12.81 77.19 + 24.16
\$ 71.09 10.73 + 38.18	\$ 30.58 46.15 + 22.30	\$ 21.47 67.26 + 10.35	\$ 32.85 34.59 + 43.25
\$ 51.66 25.76 + 31.23	\$ 20.38 36.69 + 57.07	\$ 53.80 18.57 + 21.31	\$ 70.92 19.54 + 20.16
\$ 65.71 28.46 + 13.26	\$ 12.74 25.06 + 31.42	\$ 20.52 33.26 + 64.74	\$ 36.56 10.12 + 37.28
\$ 71.03 42.52 18.79	\$ 53.80 24.08 24.35	\$ 28.03 67.41 45.97	\$ 45.82 26.95 33.36
\$ 34.58 12.37 + 56.36	\$ 56.47 34.25 + 25.32	\$ 48.23 26.01 + 32.86	\$ 34.58 63.46 + 29.39

Name ______________________________

Adding three 4-digit numbers, two or more regrouping steps

$ 34.26	$ 40.69	$ 52.05	$ 52.73
13.57	27.13	14.66	21.76
+ 52.53	+ 15.36	+ 36.82	+ 38.25

$ 38.56	$ 62.39	$ 20.08	$ 36.82
52.74	15.25	59.82	34.28
+ 13.15	+ 82.63	+ 31.32	+ 19.30

$ 93.58	$ 24.69	$ 58.32	$ 34.64
16.25	28.11	43.82	22.67
+ 91.63	+ 62.36	+ 14.16	+ 25.28

$ 32.73	$ 82.59	$ 16.75	$ 33.84
22.36	17.87	55.24	18.26
+ 37.06	+ 18.13	+ 32.18	+ 38.14

$ 63.95	$ 27.89	$ 16.29	$ 92.00
14.23	42.36	30.05	24.56
+ 25.43	+ 25.33	+ 55.82	+ 36.86

$ 29.03	$ 73.09	$ 48.35	$ 23.65
11.99	28.56	15.63	14.18
+ 62.15	+ 13.61	+ 24.18	+ 38.41

Name ______________________________

Proper placement of the decimal point and dollar sign

When you add money, there are some important things to remember.

1. When you write money, use a dollar sign ($) and decimal point (.); always be sure to have two numbers after the decimal point.

4¢	is written like this	$.04

2. When you add and subtract money, be sure to keep your columns straight. Put the dollars under dollars and the cents under cents.

Like this	\$ 12.15 + 6.01 \$ 18.16	\$ 12.15 − 6.01 \$ 6.14

Put the numbers in each box below in a column and then add them.

\$5.15 \$2.06 \$.25 =	\$9.95 \$.65 \$36.29 =
\$10.50 \$.29 \$4.25 =	\$15.21 \$.85 \$1.35 =

Name ______________________________

Solving addition story problems.

Use the menu to solve the story problems.
Answers exclude sales tax.

1. You and a friend go to the Old Fashioned Ice Cream Parlor. You each order the American Banana Split. How much will this cost? __________

2. Cindy ordered an Apple Banana Split. Ted had a Root Beer Float. How much did they owe? __________

3. Lori orders a Chocolate Super Shake with malt. Her sister has a Vanilla Super Shake with bananas and malt. How much will they pay the cashier? __________

4. Which costs more — two Super Shakes or one Grand Banana Split? ______________________

5. After the school dance, Andy and Chuck went for sodas. Chuck had a Chocolate Soda while Andy drank a Strawberry Super Soda. How much did the boys owe? __________

6. Jane has $5.00 to spend at the Old Fashioned Ice Cream Parlor. Does Jane have enough money to buy a Chocolate Soda and a Lime Freeze? __________

   How much do the two items cost? __________

Name ______________________________

Solving addition story problems.

Use the menu to solve the story problems.
All answers exclude sales tax.

---

1. After the movies, Mary took her girlfriends to the Old Fashioned Ice Cream Parlor. Mary ordered a Fancy Sundae. Debbie and Susan shared the Grand Banana Split, and Sandy had an Orange Freeze. What was the amount of their check? __________

2. John orders a regular Orange drink and Mindy has Iced Tea. Harry has Hot Chocolate, and Jim orders a small Sprite. How much is their check? __________

3. Mrs. Scott and her three sons went to the Old Fashioned Ice Cream Parlor. They each had a Pecan Banana Split. What was the amount of their check? __________

4. Angie takes two friends for ice cream. She has a Vanilla Shake. Sue has a Hot Fudge Sundae, and Chris orders a Berry Banana Split. Angie leaves the waitress a $1.00 tip. How much money does Angie spend? __________

5. At the Parlor, Jill had a Jumbo Scoop of Chocolate Chip Mint Ice Cream. Jerry had a Royal Banana Split. Tom ordered a Scotch Sundae, and Jill's mother had a cup of Sanka. What did they pay the cashier? __________

Name ______________________________

Solving addition story problems.

Use the menu to solve the story problems.
All answers exclude sales tax.

---

1. Brad and his friend shared the Giant Sundae. They each ordered a regular size Cola. How much did they owe? __________

2. What would be the cost of two Grand Banana Splits and two Rainbow Sundaes? __________

3. Sue orders the Nutty Sundae and a small Root Beer. Alice has the Special Sundae and a regular Orange drink. How much will they pay the cashier? __________

4. Kirk ordered the American Banana Split. Jack had a piece of Fudge Cake, and Robyn ordered a Fancy Sundae. They left $1.25 as a tip for the waitress. How much did they spend? __________

5. What would be the cost of three Chocolate Shakes and two small Colas? __________

6. Larry's grandmother took him to the Old Fashioned Ice Cream Parlor. He ordered a Berry Banana Split and a small Green River. His grandmother had an Apple Banana Split and a cup of coffee. How much did she spend? __________

Name ______________________________

Solving addition story problems.

Use the menu to solve the story problems.
All answers exclude sales tax.

1. Tina orders the Big Scoop Sundae with extra bananas. Lynn orders a Root Beer Float and a plain soda water. How much will this cost? __________

2. John's mother invited the fifth grade class to the Old Fashioned Ice Cream Parlor. She ordered two Rainbow Sundaes and one Fancy Sundae. How much did she spend? __________

3. Mr. Starr took his sons for banana splits. Ted had an American Banana Split, and Tony ordered a Berry Banana Split. Mr. Starr enjoyed a Pecan Banana Split. How much did Mr. Starr pay? __________

4. Becky ordered a Triple Sundae, and Steve had a Root Beer Super Float. They left a $.75 tip. How much money did they spend? __________

5. Tracy orders a Royal Banana Split. Sally has a Hot Fudge Sundae, and Jody has a Tin Roof Sundae. How much will they owe? __________

6. Find the cost of three Super Floats and two Royal Banana Splits. __________

Name ______________________________

Solving addition story problems.

Use the menu to solve the story problems.
All answers exclude sales tax.

1. Three girls and two boys went for ice cream. The boys each had a Big Scoop Sundae. The girls shared a Giant Sundae. How much was spent on ice cream? __________

2. Joanne's mother planned a birthday party for her at the Old Fashioned Ice Cream Parlor. She invited 15 girls and 10 boys to the party. They ordered three Rainbow Sundaes to be shared. How much did Joanne's mother spend on refreshments? __________

3. How much would you have to pay for an Apple Banana Split, a Hot Fudge Sundae, and a small Root Beer? __________

4. What would be the cost of two pieces of Fudge Cake and two glasses of milk? __________

5. How much would you owe if you bought four Royal Banana Splits? __________

6. Find the cost of two Giant Sundaes and three Raspberry Freezes. __________

7. How much would you have to pay for three Pecan Banana Splits and a glass of Iced Tea? __________

Name ______________________________

Figuring the cost when buying more than one of the same item.

EXAMPLE:

two Special Sundaes @ $2.45 each

$$\begin{array}{r} {}^{1} \\ \$\ 2.45 \\ +\ 2.45 \\ \hline \$\ 4.90 \end{array}$$

Find the cost of the following items.

1.	two Triple Sundaes	@	$ 3.20 each	________
2.	two Orange Freezes	@	$ 2.25 each	________
3.	three Giant Sundaes	@	$ 6.95 each	________
4.	four Grand Banana Splits	@	$ 4.95 each	________
5.	three Rainbow Sundaes	@	$12.95 each	________
6.	five regular Sprites	@	$ .95 each	________
7.	three Golden Cakes	@	$ 2.70 each	________
8.	four Tin Roof Sundaes	@	$ 2.75 each	________
9.	three Simple Sundaes	@	$ 4.00 each	________
10.	four Root Beer Floats	@	$ 2.10 each	________
11.	five Special Sundaes	@	$ 2.45 each	________
12.	three Scotch Sundaes	@	$ 3.15 each	________
13.	four Jumbo Scoops	@	$ 1.20 each	________
14.	five Hot Fudge Sundaes	@	$ 2.60 each	________

Name ______________________________

Solving addition story problems.

Solve the story problems without using the menu.

---

1. In June, the Old Fashioned Ice Cream Parlor sold 680 gallons of ice cream. During July, 529 gallons were sold, and 710 gallons were sold in August. What was the total number of gallons sold that summer? ________

2. Rich had a part-time job at the Parlor. Last year he earned $2,003.45. This year he made $1,986.04. How much did he earn in two years? ________

3. Two hundred and fifty people visited the Old Fashioned Ice Cream Parlor on Saturday. On Sunday, there were 329 people. How many people visited the Ice Cream Parlor that weekend? ________

4. In September, the Parlor sold 326 banana splits. In October, 314 were sold and during November, 298 banana splits were sold. What was the total number of banana splits sold? ________

5. Chris was a waitress at the Parlor. In June, she earned $142.95 in tips. During July, she made $125.65 and her tips in August totaled $151.25. How much did her tips total that summer? ________

6. During the winter months, the Parlor used 401 gallons of vanilla, 630 gallons of chocolate, and 806 gallons of strawberry ice cream. How many gallons were used? ________

Name ______________________________

Solving addition story problems.

Solve the story problems without using the menu.

---

1. The Parlor used 900 gallons of milk in March, 789 gallons in April, and 868 gallons in May. What was the total number of gallons used in that three-month period? ____________

2. Kevin and his twin brother, Ken, had part-time jobs at the Old Fashioned Ice Cream Parlor. In June, Kevin earned $312.68 and Ken made $317.25. Altogether, how much did the twins earn that month? ____________

3. Sandy has $233.19 in her savings account. She received a check for $203.05 from the Ice Cream Parlor. If she deposits this check in her savings account, how much will Sandy have saved? ____________

4. The Parlor used 219 pounds of almonds in June, 309 pounds in July, and 327 pounds in August. What was the total number of pounds of almonds used? ____________

5. In September, 7526 people visited the Old Fashioned Ice Cream Parlor. In October, there were 8039 customers and in November, the count was 7003. How many people ate at the ice cream parlor? ____________

6. On Wednesday, Sharon made $12.75 in tips. On Thursday, her tips totaled $14.50 and on Saturday, she received $19.05. How much did Sharon make in tips? ____________

Finding Sales Tax

Some states charge sales tax. When you buy food in a restaurant, you must pay sales tax. The amount of money you pay for sales tax depends on how much the food costs. Look carefully at the example below.

Ice Cream Parlor		
1	Nutty Sundae	3.65
1	Root Beer Float	2.10
	Subtotal	5.75
	Tax	.40
	Total	6.15

Follow the steps below:

1. Add the two items to find the subtotal.
2. Write in the tax from the Tax Table.
3. Add the subtotal and the tax to find the total.

7% Sales Tax Table

Amount of Sale	Tax	Amount of Sale	Tax
3.08 - 3.21	.22	4.79 - 4.92	.34
3.22 - 3.35	.23	4.93 - 5.07	.35
3.36 - 3.49	.24	5.08 - 5.21	.36
3.50 - 3.64	.25	5.22 - 5.35	.37
3.65 - 3.78	.26	5.36 - 5.49	.38
3.79 - 3.92	.27	5.50 - 5.64	.39
3.93 - 4.07	.28	5.65 - 5.78	.40
4.08 - 4.21	.29	5.79 - 5.92	.41
4.22 - 4.35	.30	5.93 - 6.07	.42
4.36 - 4.49	.31	6.08 - 6.21	.43
4.50 - 4.64	.32	6.22 - 6.35	.44
4.65 - 4.78	.33	6.36 - 6.49	.45

You will be practicing with a 7% sales tax table.

Find out from your teacher what tax rate your state charges.

$5.75 is between $5.65 and $5.79 on the tax table. Tax owed is $.40.

7% SALES TAX TABLE

Amount of Sale	Tax	Amount of Sale	Tax	Amount of Sale	Tax
7.50 - 7.64	.53	13.50 - 13.64	.95	19.50 - 19.64	1.37
7.65 - 7.78	.54	13.65 - 13.78	.96	19.65 - 19.78	1.38
7.79 - 7.92	.55	13.79 - 13.92	.97	19.79 - 19.92	1.39
7.93 - 8.07	.56	13.93 - 14.07	.98	19.93 - 20.07	1.40
8.08 - 8.21	.57	14.08 - 14.21	.99	20.08 - 20.21	1.41
8.22 - 8.35	.58	14.22 - 14.35	1.00	20.22 - 20.35	1.42
8.36 - 8.49	.59	14.36 - 14.49	1.01	20.36 - 20.49	1.43
8.50 - 8.64	.60	14.50 - 14.64	1.02	20.50 - 20.64	1.44
8.65 - 8.78	.61	14.65 - 14.78	1.03	20.65 - 20.78	1.45
8.79 - 8.92	.62	14.79 - 14.92	1.04	20.79 - 20.92	1.46
8.93 - 9.07	.63	14.93 - 15.07	1.05	20.93 - 21.07	1.47
9.08 - 9.21	.64	15.08 - 15.21	1.06	21.08 - 21.21	1.48
9.22 - 9.35	.65	15.22 - 15.35	1.07	21.22 - 21.35	1.49
9.36 - 9.49	.66	15.36 - 15.49	1.08	21.36 - 21.49	1.50
9.50 - 9.64	.67	15.50 - 15.64	1.09	21.50 - 21.64	1.51
9.65 - 9.78	.68	15.65 - 15.78	1.10	21.65 - 21.78	1.52
9.79 - 9.92	.69	15.79 - 15.92	1.11	21.79 - 21.92	1.53
9.93 - 10.07	.70	15.93 - 16.07	1.12	21.93 - 22.07	1.54
10.08 - 10.21	.71	16.08 - 16.21	1.13	22.08 - 22.21	1.55
10.22 - 10.35	.72	16.22 - 16.35	1.14	22.22 - 22.35	1.56
10.36 - 10.49	.73	16.36 - 16.49	1.15	22.36 - 22.49	1.57
10.50 - 10.64	.74	16.50 - 16.64	1.16	22.50 - 22.64	1.58
10.65 - 10.78	.75	16.65 - 16.78	1.17	22.65 - 22.78	1.59
10.79 - 10.92	.76	16.79 - 16.92	1.18	22.79 - 22.92	1.60
10.93 - 11.07	.77	16.93 - 17.07	1.19	22.93 - 23.07	1.61
11.08 - 11.21	.78	17.08 - 17.21	1.20	23.08 - 23.21	1.62
11.22 - 11.35	.79	17.22 - 17.35	1.21	23.22 - 23.35	1.63
11.36 - 11.49	.80	17.36 - 17.49	1.22	23.36 - 23.49	1.64
11.50 - 11.64	.81	17.50 - 17.64	1.23	23.50 - 23.64	1.65
11.65 - 11.78	.82	17.65 - 17.78	1.24	23.65 - 23.78	1.66
11.79 - 11.92	.83	17.79 - 17.92	1.25	23.79 - 23.92	1.67
11.93 - 12.07	.84	17.93 - 18.07	1.26	23.93 - 24.07	1.68
12.08 - 12.21	.85	18.08 - 18.21	1.27	24.08 - 24.21	1.69
12.22 - 12.35	.86	18.22 - 18.35	1.28	24.22 - 24.35	1.70
12.36 - 12.49	.87	18.36 - 18.49	1.29	24.36 - 24.49	1.71
12.50 - 12.64	.88	18.50 - 18.64	1.30	24.50 - 24.64	1.72
12.65 - 12.78	.89	18.65 - 18.78	1.31	24.65 - 24.78	1.73
12.79 - 12.92	.90	18.79 - 18.92	1.32	24.79 - 24.92	1.74
12.93 - 13.07	.91	18.93 - 19.07	1.33	24.93 - 25.07	1.75
13.08 - 13.21	.92	19.08 - 19.21	1.34	25.08 - 25.21	1.76
13.22 - 13.35	.93	19.22 - 19.35	1.35	25.22 - 25.35	1.77
13.36 - 13.49	.94	19.36 - 19.49	1.36	25.36 - 25.49	1.78

Name ______________________________

Using a sales tax table.

Use the tax table on page 36 to find the sales tax on the amounts shown below.

	Amount	Sales Tax	Total
1.	$ 7.75	________	________
2.	$ 8.50	________	________
3.	$ 9.10	________	________
4.	$10.35	________	________
5.	$11.15	________	________
6.	$12.05	________	________
7.	$13.77	________	________
8.	$15.48	________	________
9.	$16.75	________	________
10.	$17.98	________	________
11.	$19.92	________	________
12.	$20.95	________	________
13.	$22.80	________	________
14.	$23.05	________	________
15.	$25.25	________	________

Name ______________________

1. Add the check to find the subtotal.
2. Use the tax table to find amount of tax.
3. Add the subtotal and tax to find the total.

Ice Cream Parlor		
1	American Banana Split	$3.15
1	regular Cola	.95
1	Simple Sundae	4.00
	Subtotal	
	Tax	
	Total	

Ice Cream Parlor		
1	Golden Cake	$2.70
1	Coffee	.90
1	Giant Sundae	6.95
	Subtotal	
	Tax	
	Total	

Ice Cream Parlor		
1	Special Sundae	$2.45
1	Scotch Sundae	3.15
1	White Cake	2.70
	Subtotal	
	Tax	
	Total	

Ice Cream Parlor		
1	Vanilla Super Soda	$2.60
1	Giant Sundae	6.95
1	Grand Banana Split	4.95
	Subtotal	
	Tax	
	Total	

Name ______________________________

1. Add the check to find the subtotal.
2. Use the tax table to find amount of tax.
3. Add the subtotal and tax to find the total.

Ice Cream Parlor		
1	Nutty Sundae	$3.65
1	Chocolate Soda	2.25
1	Tin Roof Sundae	2.75
1	small Orange drink	.65
1	small Root Beer	.65
	Subtotal	
	Tax	
	Total	

Ice Cream Parlor		
1	Grand Banana Split	$4.95
1	Hot Chocolate	.75
1	Triple Sundae	3.20
1	Hot Fudge Sundae	2.60
1	Root Beer Float	2.10
	Subtotal	
	Tax	
	Total	

Ice Cream Parlor		
1	Fudge Cake	$2.80
1	Soda Water	.05
1	Simple Sundae	4.00
1	Royal Banana Split	3.00
1	Berry Banana Split	2.80
	Subtotal	
	Tax	
	Total	

Ice Cream Parlor		
1	Rainbow Sundae	$12.95
1	Tin Roof Sundae	2.75
1	Apple Banana Split	2.80
1	Scotch Sundae	3.15
	Subtotal	
	Tax	
	Total	

Name ______________________________

1. Add the check to find the subtotal.
2. Use the tax table to find amount of tax.
3. Add the subtotal and tax to find the total.

Ice Cream Parlor		
1	Golden Cake	$2.70
2	Tin Roof Sundaes	5.50
1	Berry Banana Split	2.80
3	Vanilla Shakes	6.75
	Subtotal	
	Tax	
	Total	

Ice Cream Parlor		
4	Iced Teas	$2.80
1	Grand Banana Split	4.95
1	Simple Sundae	4.00
1	Tin Roof Sundae	2.75
2	Chocolate Milks	1.60
	Subtotal	
	Tax	
	Total	

Ice Cream Parlor		
1	Special Sundae	$ 2.45
1	Rainbow Sundae	12.95
2	Hot Chocolates	1.50
1	American Banana Split	3.15
	Subtotal	
	Tax	
	Total	

Ice Cream Parlor		
2	Strawberry Sodas	$ 4.50
1	Scotch Sundae	3.15
3	Nutty Sundaes	10.95
3	Orange Freezes	6.75
	Subtotal	
	Tax	
	Total	

Name ______________________________

1. Add the check to find the subtotal.
2. Use the tax table to find amount of tax.
3. Add the subtotal and tax to find the total.

Ice Cream Parlor		
1	Root Beer Float	$ 2.10
1	Nutty Sundae	3.65
1	Fudge Cake	2.80
1	Triple Sundae	3.20
1	Rainbow Sundae	12.95
	Subtotal	
	Tax	
	Total	

Ice Cream Parlor		
1	Giant Sundae	$6.95
1	Pineapple Soda	2.25
1	Special Sundae	2.45
3	Iced Teas	2.10
	Subtotal	
	Tax	
	Total	

Ice Cream Parlor		
1	Hot Tea	$ .70
1	Golden Cake	2.70
1	Lime Freeze	2.25
1	Simple Sundae	4.00
1	Scotch Sundae	3.15
	Subtotal	
	Tax	
	Total	

Ice Cream Parlor		
1	Chocolate Shake	$2.25
1	Hot Fudge Sundae	2.60
1	Royal Banana Split	3.00
2	Iced Teas	1.40
2	small Colas	1.30
	Subtotal	
	Tax	
	Total	

Name ________________________________

1. Use the menu to find prices.
2. Add check to find subtotal.
3. Use tax table to find amount of tax.
4. Add the subtotal and tax.

Ice Cream Parlor		
2	Vanilla Super Sodas	
1	Scotch Sundae	
1	Hot Fudge Sundae	
2	Simple Sundaes	
	Subtotal	
	Tax	
	Total	

Ice Cream Parlor		
1	Special Sundae	
4	regular Colas	
2	Tin Roof Sundaes	
2	Triple Sundaes	
2	Root Beer Floats	
	Subtotal	
	Tax	
	Total	

Ice Cream Parlor		
2	White Cakes	
1	Rainbow Sundae	
1	Fancy Sundae	
1	Hot Fudge Sundae	
2	Milks	
	Subtotal	
	Tax	
	Total	

Ice Cream Parlor		
1	Apple Banana Split	
1	Chocolate Super Soda	
2	Scotch Sundaes	
4	Golden Cakes	
	Subtotal	
	Tax	
	Total	

Name ________________________________

1. Use the menu to find prices.
2. Add the check to find subtotal.
3. Use tax table to find amount of tax.
4. Add the subtotal and tax.

Ice Cream Parlor		
1	Giant Sundae	
1	Rainbow Sundae	
1	regular Cola	
1	Milk	
1	Golden Cake	
	Subtotal	
	Tax	
	Total	

Ice Cream Parlor		
1	Big Scoop Sundae	
2	Pecan Banana Splits	
2	Nutty Sundaes	
3	Raspberry Freezes	
	Subtotal	
	Tax	
	Total	

Ice Cream Parlor		
2	Simple Sundaes	
2	White Cakes	
6	Soda Waters	
1	regular Cola	
1	Grand Banana Split	
	Subtotal	
	Tax	
	Total	

Ice Cream Parlor		
4	Hot Chocolates	
3	Special Sundaes	
1	Pecan Banana Split	
2	Chocolate Shakes	
6	regular Root Beers	
	Subtotal	
	Tax	
	Total	

Name ______________________________

POST TEST — ADDITION

1. Write the amounts below using the dollar sign ($) and decimal point (.).

eight cents	________
thirty-six cents	________
fourteen dollars and six cents	________

2.	$ .62	$ .75	$ .41	$ .24	$ 1.14
	+ .21	+ .12	.28	.12	+ 1.32
			+ .33	+ .32	

3.	$ 6.42	$ 5.25	$ 3.21	$ .37	$ .56
	+ 1.50	3.11	2.14	+ .15	+ .38
		+ 1.22	+ 1.12		

4.	$ .29	$ .97	$ 4.15	$ 4.37	$ 2.77
	+ .83	+ .56	+ 4.18	+ 2.25	+ 4.33

5.	$ 6.48	$ 4.99	$ 6.92	$ 8.64
	+ 1.77	+ 9.56	+ 5.49	2.54
				+ 3.13

6.	$ 8.94	$ 35.76	$ 87.15	$ 64.71
	2.56	+ 16.77	+ 53.96	28.46
	+ 5.33			+ 13.28

Name ______________________

POST TEST — ADDITION

Solve the story problems.

---

1. Janis ordered a Triple Sundae which cost $3.20 and a regular Cola for $.95. How much did she owe? ________

2. Laurie took two friends to the Old Fashioned Ice Cream Parlor. She ordered a Chocolate Shake which cost $2.25. Sue had a Nutty Sundae for $3.65, and Larry ordered a piece of Golden Cake for $2.70. They each ordered a small Sprite which cost $.65 a glass. Laurie left a $1.50 tip for the waitress and paid $.74 tax. How much did she spend? ________

3. Find the cost of the following items:

four Giant Sundaes	@	$ 6.95 each	________
two Rainbow Sundaes	@	$12.95 each	________
three Super Sodas	@	$ 2.60 each	________
five Nutty Sundaes	@	$ 3.65 each	________

4. Earl earned $16.40 on Monday, $25.68 on Tuesday, and $52.12 on Saturday. How much did he earn altogether? ________

5. Mary earned $216.24 in tips in June, $324.40 during July, and $420.24 in August. What did her tips amount to that summer? ________

Name ______________________________

Subtracting two 2-digit numbers, no regrouping

EXAMPLE:

$$\begin{array}{r} \$\ .24 \\ -\ .11 \\ \hline \$\ .13 \end{array}$$

$ .25 − .12	$ .37 − .10	$ .95 − .14	$ .88 − .11
$ .70 − .30	$ .44 − .14	$ .36 − .14	$ .59 − .13
$ .86 − .11	$ .90 − .40	$ .88 − .52	$ .97 − .12
$ .27 − .12	$ .56 − .31	$ .36 − .10	$ .80 − .30
$ .88 − .14	$ .79 − .20	$ .24 − .12	$ .90 − .50
$ .55 − .13	$ .65 − .12	$ .81 − .31	$ .60 − .50

Name ____________________

Subtracting two 2-digit numbers, no regrouping

$ .46 − .14	$ .53 − .13	$ .22 − .20	$ .18 − .13
$ .88 − .27	$ .77 − .22	$ .98 − .46	$ .64 − .13
$ .51 − .10	$ .81 − .11	$ .36 − .12	$ .27 − .13
$ .77 − .47	$ .56 − .15	$ .68 − .15	$ .34 − .12
$ .84 − .13	$ .92 − .11	$ .29 − .12	$ .41 − .31
$ .27 − .15	$ .36 − .14	$ .29 − .10	$ .44 − .31

Name ______________________

Subtracting two 3-digit numbers, no regrouping

EXAMPLE:

$$\begin{array}{r} \$\ 4.56 \\ -\ 2.24 \\ \hline \$\ 2.32 \end{array}$$

\$ 2.25 − 1.15	\$ 4.58 − 3.24	\$ 9.50 − 5.20	\$ 9.75 − 4.34
\$ 9.76 − 5.34	\$ 6.49 − 4.16	\$ 5.88 − 2.51	\$ 8.64 − 2.12
\$ 5.85 − 2.15	\$ 3.65 − 2.12	\$ 8.25 − 1.14	\$ 5.34 − 4.21
\$ 7.38 − 4.12	\$ 8.60 − 5.20	\$ 6.25 − 4.13	\$ 3.64 − 2.13
\$ 8.75 − 5.33	\$ 4.58 − 3.23	\$ 8.25 − 1.15	\$ 4.86 − 3.24
\$ 6.49 − 5.15	\$ 6.26 − 1.16	\$ 6.90 − 5.80	\$ 2.50 − 1.20

Name ______________________

Subtracting two 3-digit numbers, no regrouping

\$ 4.66 - 2.12	\$ 5.88 - 2.55	\$ 9.05 - 2.00	\$ 9.21 - 7.00
\$ 7.06 - 4.03	\$ 4.29 - 2.04	\$ 8.08 - 1.03	\$ 2.50 - 2.30
\$ 7.05 - 3.00	\$ 8.21 - 6.00	\$ 6.08 - 5.08	\$ 3.46 - 1.30
\$ 2.46 - 2.30	\$ 6.09 - 4.02	\$ 2.35 - 1.10	\$ 8.28 - 8.04
\$ 8.50 - 2.30	\$ 5.04 - 2.03	\$ 2.05 - 1.00	\$ 2.33 - 2.03
\$ 6.35 - 1.20	\$ 5.28 - 3.04	\$ 6.35 - 2.20	\$ 3.08 - 1.02

Name ________________________

Subtracting two 4-digit numbers, no regrouping

EXAMPLE:

$$
\begin{array}{r}
\$\ 26.37 \\
-\ 15.26 \\
\hline
\$\ 11.11
\end{array}
$$

$ 12.36 - 10.11	$ 15.18 - 11.01	$ 18.39 - 15.26	$ 25.16 - 22.05
$ 24.35 - 12.14	$ 11.26 - 10.14	$ 20.49 - 10.27	$ 15.69 - 14.53
$ 31.26 - 10.15	$ 40.65 - 30.32	$ 18.25 - 17.20	$ 16.92 - 13.81
$ 16.93 - 15.81	$ 25.14 - 24.01	$ 17.45 - 15.23	$ 23.78 - 21.62
$ 14.29 - 12.18	$ 32.57 - 21.43	$ 35.18 - 31.07	$ 42.68 - 31.52
$ 28.49 - 15.27	$ 26.44 - 15.22	$ 15.95 - 12.94	$ 29.99 - 21.67

Name ______________________________

Subtracting two 4-digit numbers, no regrouping

$ 83.68 - 22.12	$ 95.84 - 22.51	$ 99.05 - 55.04	$ 88.50 - 55.20
$ 87.35 - 14.12	$ 19.50 - 15.20	$ 55.86 - 22.51	$ 42.45 - 21.14
$ 84.55 - 33.24	$ 16.58 - 14.15	$ 32.50 - 12.20	$ 85.35 - 44.12
$ 99.75 - 25.34	$ 83.64 - 12.12	$ 44.66 - 22.02	$ 86.46 - 14.16
$ 35.84 - 22.52	$ 66.33 - 44.12	$ 19.95 - 15.34	$ 75.88 - 22.51
$ 54.55 - 33.03	$ 76.06 - 44.02	$ 37.37 - 24.12	$ 88.80 - 55.50

Name ______________________________

Subtracting two 2-digit numbers, one-step regrouping

EXAMPLE:

```
     3 1
  $ .46
  - .28
  -----
  $ .18
```

$ .47 − .29	$ .93 − .85	$ .87 − .29	$ .66 − .49
$ .57 − .49	$ .93 − .66	$ .42 − .19	$ .82 − .75
$ .95 − .46	$ .70 − .08	$ .53 − .05	$ .72 − .56
$ .72 − .65	$ .60 − .04	$ .83 − .25	$ .44 − .38
$ .63 − .25	$ .55 − .38	$ .97 − .29	$ .67 − .09
$ .87 − .19	$ .63 − .07	$ .22 − .15	$ .26 − .19

Name ______________________________

Subtracting two 2-digit numbers, one-step regrouping

\$ .85 – .36	\$ .76 – .28	\$ .72 – .39	\$ .83 – .76
\$ .70 – .29	\$ .66 – .47	\$ .40 – .28	\$ .30 – .19
\$ .23 – .09	\$ .90 – .18	\$ .55 – .28	\$ .25 – .09
\$ .86 – .68	\$ .25 – .18	\$ .22 – .05	\$ .33 – .14
\$ .93 – .67	\$ .42 – .38	\$ .86 – .38	\$ .46 – .17
\$ .22 – .14	\$ .76 – .49	\$ .85 – .06	\$ .77 – .08

Name ______________________________

Subtracting two 3-digit numbers, two or more regrouping steps

EXAMPLE:

$ 4.26 (regrouped: 3 11 1 written above, 4 and 2 crossed out)
\- 1.78
$ 2.48

$ 9.22 - 4.63	$ 8.17 - 5.29	$ 3.64 - 1.95	$ 2.42 - 1.68
$ 6.21 - 4.42	$ 3.73 - 1.95	$ 4.27 - 1.88	$ 9.50 - 4.94
$ 8.33 - 7.54	$ 3.52 - 1.94	$ 3.74 - 1.98	$ 5.42 - 1.54
$ 6.22 - 2.68	$ 3.28 - 1.89	$ 6.51 - 1.84	$ 2.21 - 1.42
$ 8.24 - 4.68	$ 9.21 - 4.43	$ 6.52 - 2.87	$ 8.70 - 2.95
$ 4.72 - 1.96	$ 2.32 - 1.54	$ 6.26 - 5.89	$ 2.87 - 1.99

Name ______________________________

Subtracting two 3-digit numbers, two or more regrouping steps

\$ 6.67 - 2.88	\$ 7.45 - 6.57	\$ 8.25 - 5.29	\$ 6.22 - 4.44
\$ 9.32 - 4.53	\$ 9.40 - 2.59	\$ 2.74 - 1.95	\$ 3.53 - 1.54
\$ 7.21 - 4.52	\$ 3.84 - 2.95	\$ 3.81 - 1.95	\$ 2.52 - 1.93
\$ 6.40 - 2.54	\$ 3.50 - 1.93	\$ 5.27 - 1.88	\$ 8.22 - 4.63
\$ 5.07 - 2.68	\$ 3.27 - 2.88	\$ 4.40 - 1.69	\$ 5.21 - 4.42
\$ 3.62 - 1.84	\$ 8.21 - 4.63	\$ 3.32 - 1.94	\$ 5.40 - 2.54

Name ______________________________

Subtracting two 4-digit numbers, two or more regrouping steps

EXAMPLE:

5 14 17 1
$ 65.80
- 36.93
$ 28.87

$ 65.86 - 29.97	$ 36.47 - 27.58	$ 98.43 - 89.59	$ 65.80 - 37.94
$ 52.81 - 43.95	$ 84.60 - 56.74	$ 76.52 - 49.69	$ 52.87 - 43.99
$ 32.43 - 18.79	$ 57.32 - 49.87	$ 85.36 - 76.49	$ 88.40 - 79.59
$ 94.86 - 29.98	$ 84.83 - 45.94	$ 74.32 - 15.95	$ 88.32 - 79.36
$ 24.70 - 19.87	$ 73.82 - 44.97	$ 72.32 - 15.96	$ 97.21 - 58.85
$ 32.48 - 23.59	$ 85.32 - 66.44	$ 84.24 - 36.88	$ 28.97 - 19.98

Name ______________________________

Subtracting two 4-digit numbers, two or more regrouping steps

\$ 82.43 - 46.46	\$ 97.18 - 88.29	\$ 55.06 - 26.98	\$ 46.47 - 17.58
\$ 22.42 - 19.86	\$ 24.16 - 18.78	\$ 42.81 - 41.95	\$ 84.93 - 56.74
\$ 55.46 - 46.37	\$ 38.40 - 19.36	\$ 32.46 - 19.79	\$ 67.30 - 49.89
\$ 58.47 - 49.58	\$ 22.57 - 19.86	\$ 95.01 - 28.97	\$ 46.47 - 27.58
\$ 82.35 - 46.46	\$ 36.53 - 19.59	\$ 32.81 - 17.79	\$ 54.67 - 26.98
\$ 63.40 - 54.49	\$ 72.28 - 66.49	\$ 26.43 - 24.59	\$ 24.63 - 16.74

Name ______________________________

Subtracting
Zero difficulty

EXAMPLE:

```
  7 9₁
$ 8.00
- 2.81
------
$ 5.19
```

$ 3.00 - 1.88	$ 6.00 - 4.84	$ 4.00 - .59	$ 6.00 - .32
$ 6.00 - 2.29	$ 5.00 - 1.28	$ 1.00 - .29	$ 2.00 - 1.94
$ 1.00 - .99	$ 5.00 - 1.89	$ 9.00 - 4.12	$ 4.00 - 3.19
$ 7.00 - 4.19	$ 8.00 - 3.66	$ 6.00 - 2.58	$ 3.00 - 1.93
$ 5.00 - 4.12	$ 6.00 - 1.16	$ 6.00 - 2.98	$ 5.00 - 4.94
$ 5.00 - 4.46	$ 3.00 - .83	$ 8.00 - 2.82	$ 5.00 - .69

Name ______________________________

Subtracting
Zero Difficulty

EXAMPLE:

```
   3 9 9 1
$  40.00
-   2.98
--------
$  37.02
```

$ 60.00 − 1.82	$ 20.00 − .89	$ 80.00 − .65	$ 20.00 − 5.55
$ 10.00 − 9.95	$ 30.00 − 6.56	$ 70.00 − 2.25	$ 60.00 − 1.95
$ 40.00 − 2.89	$ 50.00 − .46	$ 40.00 − 4.12	$ 30.00 − 1.98
$ 20.00 − 2.89	$ 10.00 − .46	$ 60.00 − 4.12	$ 10.00 − 1.98
$ 20.00 − 1.45	$ 60.00 − .98	$ 30.00 − .84	$ 20.00 − 9.99
$ 30.00 − 1.89	$ 60.00 − 4.44	$ 30.00 − 4.18	$ 80.00 − 2.98

Name ______________________________

Subtracting
Zero difficulty

\$ 20.00 - 15.99	\$ 70.00 - 19.95	\$ 40.00 - 27.82	\$ 50.00 - 18.97
\$ 60.00 - 42.84	\$ 30.00 - 16.93	\$ 20.00 - 9.99	\$ 40.00 - 12.36
\$ 80.00 - 15.98	\$ 20.00 - 14.98	\$ 60.00 - 15.35	\$ 40.00 - 18.91
\$ 70.00 - 15.42	\$ 90.00 - 27.99	\$ 50.00 - 44.95	\$ 70.00 - 27.97
\$ 40.00 - 33.91	\$ 50.00 - 25.55	\$ 30.00 - 18.98	\$ 40.00 - 22.12
\$ 30.00 - 28.72	\$ 20.00 - 18.88	\$ 60.00 - 21.36	\$ 50.00 - 19.39

Name ______________________________

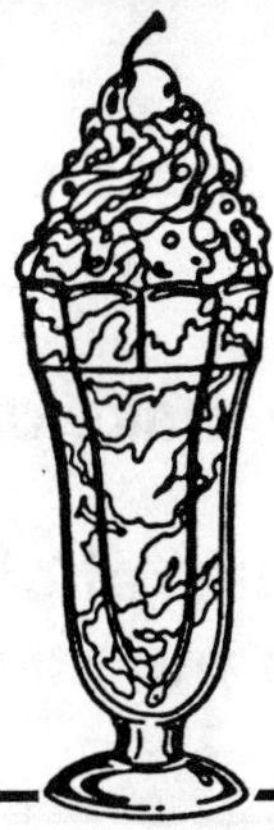

Subtracting
Zero difficulty

\$ 20.00 - 13.91	\$ 40.00 - 21.33	\$ 60.00 - 14.82	\$ 30.00 - 22.13
\$ 50.00 - 11.81	\$ 30.00 - 26.22	\$ 40.00 - 33.17	\$ 50.00 - 42.36
\$ 60.00 - 11.77	\$ 20.00 - 10.99	\$ 60.00 - 28.16	\$ 70.00 - 29.20
\$ 70.00 - 14.22	\$ 30.00 - 15.21	\$ 80.00 - 52.17	\$ 90.00 - 37.12
\$ 50.00 - 12.32	\$ 80.00 - 26.11	\$ 70.00 - 42.34	\$ 40.00 - 31.88
\$ 70.00 - 49.21	\$ 40.00 - 25.11	\$ 70.00 - 36.28	\$ 50.00 - 27.33

Name ____________________________

Subtracting to make change.

Write the amount of change due from each purchse.

	COST	PAID	CHANGE
1.	$ .26	$ .30	________
2.	$ .14	$ .50	________
3.	$ .76	$ 1.00	________
4.	$ .68	$ .75	________
5.	$ .72	$ 1.00	________
6.	$ 1.03	$ 2.00	________
7.	$ 2.53	$ 3.00	________
8.	$ 6.29	$ 10.00	________
9.	$ 3.98	$ 5.00	________
10.	$ 8.35	$ 9.00	________
11.	$ 14.50	$ 20.00	________
12.	$ 12.50	$ 15.00	________
13.	$ 38.18	$ 40.00	________
14.	$ 46.80	$ 50.00	________
15.	$ 55.45	$ 60.00	________
16.	$ 71.92	$ 75.00	________

Name ____________________

Solving subtraction story problems.

Use the menu to solve the story problems.
Answers exclude sales tax.

---

1. Scott took $5.00 to the Old Fashioned Ice Cream Parlor. He bought a Jumbo Scoop of Strawberry Ice Cream. How much change did he receive? ________

2. How much more would a Grand Banana Split cost than a Hot Fudge Sundae? ________

3. How much more would a Berry Banana Split cost than a Special Sundae? ________

4. What is the difference between the cost of a Tin Roof Sundae and a Scotch Sundae? ________

5. Jim ordered a piece of Fudge Cake, and Mike ordered a Big Scoop Sundae. How much more will Mike pay than Jim? ________

6. Andy had $5.00 to spend at the Old Fashioned Ice Cream Parlor. He ordered the American Banana Split. How much change will he receive? ________

7. Sam has $3.50. He wants to buy a Giant Sundae. How much more money does he need? ________

Name ______________________________

Solving subtraction story problems.

Use the menu to solve the story problems.
Answers exclude sales tax.

---

1. Richard and Tom want to share a Giant Sundae. They have only $3.56. How much more money will they need? ____________

2. Sherri ordered a Fancy Sundae, and Ann had a Triple Sundae. How much more did Ann spend than Sherri? ____________

3. Four boys want to share a Rainbow Sundae. They have only $10.89. How much more money will they need? ____________

4. Jack took $10.00 to the Old Fashioned Ice Cream Parlor. He bought a Simple Sundae. How much money does he have left? ____________

5. Debbie wanted to order a Nutty Sundae. She had only $2.86. How much more money would she need? ____________

6. Eric had a Grand Banana Split at the Old Fashioned Ice Cream Parlor. He used a $20.00 bill to pay his check. How much change did he get back? ____________

7. Which costs more, the Fancy Sundae or the Pecan Banana Split? ____________________

   How much more? ____________

Name ______________________________

Solving addition and subtraction story problems.

Use the menu to solve the story problems.
Answers exclude sales tax.

---

1. Linda's mother gave her $8.00 to take a friend to the Ice Cream Parlor. Linda ordered the Triple Sundae and her friend ordered an Apple Banana Split. How much change did Linda receive? __________

2. June ordered a Strawberry Super Soda, and Jill had a Scotch Sundae. How much more did Jill pay than June? __________

3. Steve and Jerry took $10.00 to the Old Fashioned Ice Cream Parlor. They spent $4.25. How much change did the cashier give them? __________

4. Eight girls went to the Ice Cream Parlor. They ordered the Rainbow Sundae and paid for it with a $20.00 bill. How much change did they get back? __________

5. Todd's sundae cost $3.65, and Jack's sundae was $4.00. How much more did Jack spend than Todd? __________

6. Mr. Smith took his family to the Ice Cream Parlor. They ordered a Nutty Sundae, a piece of White Cake, and a Tin Roof Sundae. They also ordered four small Colas. How much change did Mr. Smith get back from $15.00? __________

Name ______________________________

Solving addition and subtraction story problems.

Use the menu to solve the story problems.
Answers exclude sales tax.

---

1. Steve had an Orange Freeze, and Randy ordered a Grand Banana Split. How much was their bill? __________

   How much change will they get back from $10.00? __________

2. You order two Vanilla Sodas and a piece of Fudge Cake. How much does this cost? __________

   How much change will you get back from $10.00? __________

3. Which costs more, the Special Sundae or the Hot Fudge Sundae? ________________

   How much more? __________

4. You order a Grand Banana Split and a Scotch Sundae. How much will you owe? __________

   How much should you get from $20.00? __________

5. Tony had Fudge Cake which costs $2.80. Sharon ordered a Grand Banana Split for $4.95. How much was their bill? __________

   How much change did they get from $10.00? __________

6. John and Dave went to the Old Fashioned Ice Cream Parlor. John ordered an American Banana Split, and Dave had a Fancy Sundae.
   Which boy spent more money? ________________

   How much more? __________

Name ______________________________

Solving addition and subtraction story problems.

Use the menu to solve the story problems.
Answers exclude sales tax.

---

1. Mrs. Davis took her class to the Old Fashioned Ice Cream Parlor. She ordered two Rainbow Sundaes. How much did this cost? __________

   How much change should she get from $30.00? __________

2. Jeff and Larry went to the Ice Cream Parlor. Jeff ordered the Royal Banana Split, and Larry had a piece of Fudge Cake and a glass of milk. Which boy spent more? __________

   How much more? __________

3. Mary ordered a Triple Sundae, and Lisa had a piece of Fudge Cake and a glass of milk. Which girl spent more money? __________

   How much more? __________

4. You and a friend each order a piece of Golden Cake and two regular Root Beers. How much will your check be? __________

   How much change will you receive from a $10.00 bill? __________

5. If David had saved $15.00 to take his family to the Ice Cream Parlor, how much more will he need to save to have enough money to pay for two Rainbow Sundaes? __________

6. Three boys each order an Orange Freeze. They give the cashier a $20.00 bill. How much change will they receive? __________

Name ______________________________

Solving addition and subtraction story problems.

Solve the following story problems without using the menu.

1. The Parlor used 869 pounds of nuts in 1989 and 905 pounds during 1990. How many more pounds of nuts were used in 1990? __________

2. In March, the Parlor used 38 jars of cherries and in April, 45 jars were used. How many more jars were used in April than in March? __________

3. Sandy and Sally were waitresses at the Parlor. One week Sandy earned $42.25 in tips, and Sally earned $38.73. How much more did Sandy earn than Sally? __________

4. Mark needed $115.25 for a bike he wanted to buy. He got a job at the Old Fashioned Ice Cream Parlor and earned $74.50. How much more money will he need to buy the bike? __________

5. Nancy worked at the Parlor. She saved $29.50 for school clothes. She needs $110.00. How much more will she need to earn to have $110.00? __________

6. Jack earned $215.00 working at the Parlor. He spent $56.20 and put $25.00 in his savings account. How much cash did Jack have left? __________

Name ______________________________

1. Add check to find subtotal.
2. Use tax table to find amount of tax.
3. Add the subtotal and tax.
4. Subtract total from amount paid.

Ice Cream Parlor		
1	Rainbow Sundae	$12.95
3	small Sprites	1.95
2	Special Sundaes	4.90
	Subtotal	
	Tax	
	Total	

Amount paid $25.00 **Change:** __________

Ice Cream Parlor		
6	regular Colas	$ 5.70
3	Special Sundaes	7.35
1	Coffee	.90
	Subtotal	
	Tax	
	Total	

Amount paid $15.00 **Change:** __________

Ice Cream Parlor		
1	Rainbow Sundae	$12.95
1	Grand Banana Split	4.95
2	Chocolate Sodas	4.50
	Subtotal	
	Tax	
	Total	

Amount paid $25.00 **Change:** __________

Ice Cream Parlor		
2	Pecan Banana Splits	$ 5.60
1	Triple Sundae	3.20
2	Lime Freezes	4.50
	Subtotal	
	Tax	
	Total	

Amount paid $20.00 **Change:** __________

Name ______________________________

1. Add check to find subtotal.
2. Use tax table to find amount of tax.
3. Add the subtotal and tax.
4. Subtract total from amount paid.

Ice Cream Parlor		
2	Root Beer Floats	$ 4.20
1	Strawberry Soda	2.25
2	Orange Freezes	4.50
1	Fudge Cake	2.80
	Subtotal	
	Tax	
	Total	

Amount paid $15.00 **Change:** __________

Ice Cream Parlor		
1	Simple Sundae	$ 4.00
1	Grand Banana Split	4.95
1	Royal Banana Split	3.00
2	small Colas	1.30
	Subtotal	
	Tax	
	Total	

Amount paid $20.00 **Change:** __________

Ice Cream Parlor		
3	Vanilla Super Sodas	$ 7.80
2	Fancy Sundaes	5.20
1	Special Sundae	2.45
	Subtotal	
	Tax	
	Total	

Amount paid $20.00 **Change:** __________

Ice Cream Parlor		
2	Scotch Sundaes	$ 6.30
1	Nutty Sundae	3.65
1	Cola Float	2.10
	Subtotal	
	Tax	
	Total	

Amount paid $15.00 **Change:** __________

Name ______________________________

1. Add check to find subtotal.
2. Use tax table to find amount of tax.
3. Add the subtotal and tax.
4. Subtract total from amount paid.

Ice Cream Parlor		
1	Simple Sundae	$ 4.00
1	Chocolate Soda	2.25
1	regular Orange	.95
1	Orange Freeze	2.25
	Subtotal	
	Tax	
	Total	

Amount paid $15.00 **Change:** __________

Ice Cream Parlor		
1	Pecan Banana Split	$ 2.80
1	Grand Banana Split	4.95
1	Giant Sundae	6.95
2	small Colas	1.30
	Subtotal	
	Tax	
	Total	

Amount paid $20.00 **Change:** __________

Ice Cream Parlor		
1	White Cake	$ 2.70
1	Giant Sundae	6.95
1	Hot Fudge Sundae	2.60
	Subtotal	
	Tax	
	Total	

Amount paid $15.00 **Change:** __________

Ice Cream Parlor		
2	small Sprites	$ 1.30
1	Fudge Cake	2.80
1	Rainbow Sundae	12.95
	Subtotal	
	Tax	
	Total	

Amount paid $20.00 **Change:** __________

Name ______________________________

1. Add check to find subtotal.
2. Use tax table to find amount of tax.
3. Add the subtotal and tax.
4. Subtract total from amount paid.

Ice Cream Parlor		
1	regular Cola	$ .95
1	Golden Cake	2.70
1	Milk	.80
1	Big Scoop Sundae	3.05
	Subtotal	
	Tax	
	Total	

Amount paid $10.00 **Change:** __________

Ice Cream Parlor		
2	Vanilla Super Sodas	$ 5.20
3	Raspberry Freezes	6.75
2	small Colas	1.30
	Subtotal	
	Tax	
	Total	

Amount paid $20.00 **Change:** __________

Ice Cream Parlor		
2	Pecan Banana Splits	$ 5.60
2	Berry Banana Splits	5.60
2	Golden Cakes	5.40
	Subtotal	
	Tax	
	Total	

Amount paid $20.00 **Change:** __________

Ice Cream Parlor		
2	Special Sundaes	$ 4.90
2	Grand Banana Splits	9.90
1	Coffee	.90
	Subtotal	
	Tax	
	Total	

Amount paid $20.00 **Change:** __________

Name ______________________________

1. Use the menu to find prices.
2. Add check to find subtotal.
3. Use tax table to find amount of tax.
4. Add the subtotal and tax.
5. Subtract total from amount paid to find change.

Ice Cream Parlor		
1	Fancy Sundae	
2	Berry Banana Splits	
1	Scotch Sundae	
1	Fudge Cake	
	Subtotal	
	Tax	
	Total	

Amount paid $20.00 **Change:** __________

Ice Cream Parlor		
1	Chocolate Shake	
1	Nutty Sundae	
2	Simple Sundaes	
3	Chocolate Milks	
	Subtotal	
	Tax	
	Total	

Amount paid $20.00 **Change:** __________

Ice Cream Parlor		
3	Special Sundaes	
2	Pecan Banana Splits	
1	Giant Sundae	
1	Big Scoop Sundae	
	Subtotal	
	Tax	
	Total	

Amount paid $30.00 **Change:** __________

Ice Cream Parlor		
2	Hot Fudge Sundaes	
2	Apple Banana Splits	
2	American Banana Splits	
6	small Colas	
	Subtotal	
	Tax	
	Total	

Amount paid $40.00 **Change:** __________

Name ______________________________

1. Use the menu to find prices.
2. Add check to find subtotal.
3. Use tax table to find amount of tax.
4. Add the subtotal and tax.
5. Subtract total from amount paid to find change.

Ice Cream Parlor		
1	Scotch Sundae	
2	Golden Cakes	
3	Orange Freezes	
1	Giant Sundae	
	Subtotal	
	Tax	
	Total	

Amount paid $25.00 **Change:** __________

Ice Cream Parlor		
2	Big Scoop Sundaes	
1	Triple Sundae	
3	Vanilla Shakes	
2	Simple Sundaes	
	Subtotal	
	Tax	
	Total	

Amount paid $30.00 **Change:** __________

Ice Cream Parlor		
3	Iced Teas	
2	Raspberry Freezes	
2	Nutty Sundaes	
1	regular Sprite	
	Subtotal	
	Tax	
	Total	

Amount paid $20.00 **Change:** __________

Ice Cream Parlor		
2	Fancy Sundaes	
1	Pecan Banana Split	
3	Milks	
	Subtotal	
	Tax	
	Total	

Amount paid $15.00 **Change:** __________

Name ________________________________

POST TEST — ADDITION AND SUBTRACTION

1. John ordered a Nutty Sundae which sells for $3.65 and a regular Sprite for $.95. How much was his bill? (Do not include sales tax.) ________

2.	$ .62 + .22	$ .43 .25 + .31	$ 2.15 + 2.32	$ 5.20 2.26 + 2.13
3.	$ .38 + .82	$ 4.25 + 4.17	$ 2.88 + 6.33	$ 6.68 + 5.46
4.	$ 9.64 1.53 + 3.16	$38.78 +15.75	$62.75 25.46 +13.46	$ .45 – .13
5.	$35.47 –27.58	$ 4.00 – 1.88	$50.00 – 9.95	$90.00 –47.52
6.	$58.63 –29.54	$80.00 –12.95	$83.62 –29.73	$ 7.00 – 3.66

Name ______________________________

POST TEST — ADDITION AND SUBTRACTION

Solve the story problems.
Answers exclude sales tax.

---

1. Find the cost of the following items:

three Triple Sundaes	@	$ 3.20 each	________
five Chocolate Shakes	@	$ 2.25 each	________

2. Betty earned $514.24 in June, $624.50 in July, and $392.52 in August. What was the total amount she earned while working at the Ice Cream Parlor? ________

3. Howard took $10.00 to the Old Fashioned Ice Cream Parlor. He spent $2.80 on ice cream. How much change will he receive from the cashier? ________

4. Andy has $3.50 to spend on ice cream. He wants to buy a Giant Sundae which costs $6.95. How much more money does he need? ________

5. You and a friend each order a Simple Sundae. The sundaes sell for $4.00 each. How much will you pay the cashier? ________

   How much change will you get back from a $20.00 bill? ________

6. Glenn earned $265.00 working at the Old Fashioned Ice Cream Parlor. He spent $151.31 on school clothes. How much did Glenn have left from his earnings? ________

7. Barry ordered five Rainbow Sundaes which cost $12.95 apiece. He gave the cashier $70.00. How much change did Barry receive? ________

Ice Cream Parlor, Book 1 — Answer Key

PAGE 1:

1) $ .03
2) $ .07
3) $ .12
4) $ .17
5) $ .35
6) $ .60
7) $ .99
8) $ 1.14
9) $12.22
10) $30.29
11) $45.16
12) $52.02
13) $60.45
14) $75.20

PAGE 2:

$ .16	$ .17	$ .37	$ .24
$ .45	$ .36	$ .59	$ .43
$ .35	$ .38	$ .66	$ .80
$ .60	$ .39	$ .79	$ .75
$ .49	$ .65	$ .97	$ .68
$ .35	$ .31	$ .94	$ .98

PAGE 3:

$ .79	$ .84	$ .55	$ .79
$ .78	$ .96	$ .78	$ .79
$ .75	$ .88	$ .84	$ .82
$ .95	$ .97	$ .96	$ .94
$ .66	$ .57	$ .87	$ .44
$ .94	$ .96	$ .86	$ .55

PAGE 4:

$ .77	$ .99	$ .69	$ .64
$ .57	$ .98	$ .89	$ .81
$ .75	$ .79	$ .89	$ .76
$ .56	$ .39	$ .89	$ .69
$ .64	$ .89	$ .86	$ .85
$ .75	$ .99	$ .97	$ .67

PAGE 5:

$ .79	$ .77	$ .69	$ .79
$ .57	$ .48	$ .77	$ .67
$ .56	$ .68	$ .84	$ .79
$ .79	$ .49	$ .88	$ .97
$ .59	$ .39	$ .59	$ .78
$ .68	$ .69	$ .89	$ .94

PAGE 6:

$ 2.53	$ 2.45	$ 5.19	$ 4.50
$ 5.17	$ 3.39	$ 6.18	$ 7.27
$ 8.73	$ 4.27	$ 6.06	$ 8.05
$ 6.61	$ 8.81	$ 8.72	$ 8.58
$ 8.73	$ 5.09	$ 9.81	$ 7.70
$ 7.15	$ 2.55	$ 9.08	$ 9.70

PAGE 7:

$ 5.85	$ 6.93	$ 8.66	$ 9.71
$ 7.65	$ 9.97	$ 7.82	$ 9.93
$ 9.93	$ 9.86	$ 6.45	$ 9.56
$ 8.46	$ 9.86	$ 7.69	$ 6.48
$ 6.65	$ 7.44	$ 6.88	$ 9.87
$ 7.87	$ 9.96	$ 7.99	$ 9.97

PAGE 8:

$ 9.94	$ 9.86	$ 9.61	$ 9.73
$ 8.39	$ 9.85	$ 9.96	$ 4.97
$ 6.59	$ 7.45	$ 5.73	$ 7.38
$ 7.85	$ 9.97	$ 9.65	$ 8.78
$ 8.67	$ 9.82	$ 9.29	$ 8.18
$ 7.86	$ 9.95	$ 9.19	$ 9.19

PAGE 9:

$ .50	$ .70	$ .90	$ .63
$ .40	$ .60	$ .82	$ .90
$ .70	$ .61	$ .87	$ .90
$ .44	$ .63	$ .95	$ .52
$ .60	$ .54	$ .40	$ .51
$ .44	$ .88	$ .43	$ .62

PAGE 10:

$ .30	$ .30	$ .34	$ .43
$ .31	$ .40	$ .63	$ .90
$ .41	$ .32	$ .81	$ .60
$ .31	$ .30	$ .92	$ .50
$ .40	$ .43	$ .90	$ .90
$ .60	$ .34	$ .51	$ .41

PAGE 11:

$ 1.10	$ 1.41	$ 1.60	$ 1.04
$ 1.00	$ 1.85	$ 1.01	$ 1.13
$ 1.12	$ 1.30	$ 1.70	$ 1.22
$ 1.10	$ 1.52	$ 1.82	$ 1.90
$ 1.20	$ 1.20	$ 1.13	$ 1.80
$ 1.04	$ 1.13	$ 1.44	$ 1.50

PAGE 12:

$ 1.82	$ 1.12	$ 1.60	$ 1.57
$ 1.61	$ 1.73	$ 1.15	$ 1.70
$ 1.72	$ 1.57	$ 1.73	$ 1.12
$ 1.21	$ 1.60	$ 1.11	$ 1.33
$ 1.37	$ 1.08	$ 1.70	$ 1.16
$ 1.24	$ 1.43	$ 1.21	$ 1.10

PAGE 13:

$ 1.83	$ 1.61	$ 1.12	$ 1.22
$ 1.10	$ 1.20	$ 1.03	$ 1.01
$ 1.44	$ 1.11	$ 1.24	$ 1.12
$ 1.21	$ 1.31	$ 1.30	$ 1.31
$ 1.32	$ 1.42	$ 1.32	$ 1.76
$ 1.35	$ 1.45	$ 1.98	$ 1.60

PAGE 14:

$ 2.31	$ 2.60	$ 8.30	$ 8.70
$ 5.30	$ 6.70	$ 8.62	$ 9.91
$ 8.80	$ 9.83	$ 2.33	$ 2.61
$ 9.90	$ 9.61	$ 5.20	$ 6.70
$ 2.60	$ 4.40	$ 7.60	$ 8.40
$ 6.33	$ 8.30	$ 7.60	$ 4.70

PAGE 15:

$ 6.63	$ 6.86	$ 8.27	$ 8.90
$ 9.80	$ 8.37	$ 8.32	$ 9.38
$ 9.10	$ 9.28	$ 8.70	$ 7.67
$ 9.70	$ 9.71	$ 8.33	$ 9.41
$ 8.31	$ 9.17	$ 9.54	$ 9.30
$ 9.96	$ 7.32	$ 9.35	$ 9.40

PAGE 16:

$ 4.02	$ 6.30	$ 7.23	$ 8.18
$ 4.17	$ 5.54	$ 4.80	$ 4.06
$ 4.45	$ 5.10	$ 9.04	$ 6.10
$ 7.60	$ 5.41	$ 9.01	$ 9.13
$ 7.51	$ 9.16	$ 5.10	$ 6.50
$ 7.03	$ 9.01	$ 7.05	$ 8.27

PAGE 17:

$ 9.15	$ 5.00	$ 9.50	$ 7.31
$ 6.08	$ 8.07	$ 5.21	$ 9.28
$ 4.96	$ 7.67	$ 5.27	$ 7.52
$ 9.83	$ 9.13	$ 9.22	$ 8.54
$ 6.75	$ 9.41	$ 5.21	$ 7.03
$ 9.90	$ 8.65	$ 9.05	$ 9.00

PAGE 18:

$ 15.42	$ 13.65	$ 14.08	$ 15.54
$ 14.24	$ 11.44	$ 14.10	$ 16.28
$ 14.25	$ 16.06	$ 15.05	$ 14.00
$ 18.02	$ 11.03	$ 18.85	$ 15.63
$ 11.00	$ 14.37	$ 11.01	$ 14.00
$ 13.65	$ 13.72	$ 13.34	$ 12.11

PAGE 19:

$ 11.21	$ 13.45	$ 12.01	$ 15.22
$ 18.22	$ 12.14	$ 13.20	$ 14.34
$ 16.44	$ 13.00	$ 18.01	$ 12.33
$ 14.08	$ 17.82	$ 12.20	$ 11.95
$ 14.54	$ 11.10	$ 11.00	$ 10.31
$ 16.65	$ 14.11	$ 12.21	$ 14.43

PAGE 20:

$ 50.53	$ 70.50	$108.10	$ 71.05
$ 85.22	$ 56.02	$ 42.40	$ 50.60
$ 90.11	$ 98.52	$ 25.26	$129.30
$131.10	$ 40.06	$ 81.60	$ 48.13
$ 93.35	$ 44.25	$ 53.50	$ 88.40
$ 71.20	$ 66.50	$ 65.05	$ 81.70

PAGE 21:

$ 55.63	$ 53.14	$ 80.47	$ 87.12
$108.15	$ 96.80	$106.45	$ 40.12
$141.11	$118.10	$113.16	$133.55
$123.25	$161.50	$117.25	$111.11
$141.07	$132.12	$178.11	$108.83
$151.11	$ 70.74	$151.10	$118.01

PAGE 22:

$ 10.79	$ 10.64	$ 10.14	$ 11.10
$ 10.20	$ 11.60	$ 10.81	$ 13.11
$ 11.42	$ 10.60	$ 14.43	$ 13.65
$ 11.07	$ 12.41	$ 11.14	$ 12.26
$ 10.11	$ 10.19	$ 12.19	$ 13.01
$ 10.21	$ 11.08	$ 16.51	$ 10.93

PAGE 23:

$ 10.52	$ 12.30	$ 11.26	$ 12.62
$ 13.40	$ 11.21	$ 14.00	$ 16.73
$ 10.21	$ 12.23	$ 13.13	$ 11.60
$ 13.70	$ 11.23	$ 10.20	$ 13.18
$ 12.21	$ 11.10	$ 11.22	$ 10.31
$ 11.54	$ 12.54	$ 12.03	$ 11.39

PAGE 24:

$ 85.05	$103.16	$112.35	$114.16
$120.00	$ 99.03	$ 99.08	$110.69
$108.65	$114.14	$ 93.68	$110.62
$107.43	$ 69.22	$118.52	$ 83.96
$132.34	$102.23	$141.41	$106.13
$103.31	$116.04	$107.10	$127.43

PAGE 25:

$100.36	$ 83.18	$103.53	$112.74
$104.45	$160.27	$111.22	$ 90.40
$201.46	$115.16	$116.30	$ 82.59
$ 92.15	$118.59	$104.17	$ 90.24
$103.61	$ 95.58	$102.16	$153.42
$103.17	$115.26	$ 88.16	$ 76.24

PAGE 26: $ 7.46 $ 46.89
$ 15.04 $ 17.41

PAGE 27: 1) $ 6.30 2) $ 4.90 3) $ 6.00
4) two Super Shakes 5) $ 4.85
6) yes, $ 4.50

PAGE 28: 1) $ 9.80 2) $ 3.05 3) $ 11.20
4) $ 8.65 5) $ 8.25

PAGE 29: 1) $ 8.85 2) $ 35.80 3) $ 7.70
4) $ 9.80 5) $ 8.05 6) $ 7.15

PAGE 30: 1) $ 5.30 2) $ 28.50 3) $ 8.75
4) $ 6.45 5) $ 8.35 6) $ 13.50

PAGE 31: 1) $ 13.05 2) $ 38.85 3) $ 6.05
4) $ 7.20 5) $ 12.00 6) $ 20.65
7) $ 9.10

PAGE 32:

1) $ 6.40
2) $ 4.50
3) $20.85
4) $19.80
5) $38.85
6) $ 4.75
7) $ 8.10
8) $11.00
9) $12.00
10) $ 8.40
11) $12.25
12) $ 9.45
13) $ 4.80
14) $13.00

PAGE 33: 1) 1919 2) $3989.49 3) 579
4) 938 5) $419.85 6) 1837

PAGE 34: 1) 2557 2) $629.93 3) $436.24
4) 855 5) 22,568 6) $ 46.30

PAGE 37:

1) .54/ 8.29
2) .60/ 9.10
3) .64/ 9.74
4) .72/11.07
5) .78/11.93
6) .84/12.89
7) .96/14.73
8) 1.08/16.56
9) 1.17/17.92
10) 1.26/19.24
11) 1.39/21.31
12) 1.47/22.42
13) 1.60/24.40
14) 1.61/24.66
15) 1.77/27.02

Ice Cream Parlor, Book 1 — Answer Key

PAGE 38: 8.10/.57/8.67 10.55/.74/11.29
8.30/.58/8.88 14.50/1.02/15.52

PAGE 39: 9.95/.70/10.65 13.60/.95/14.55
12.65/.89/13.54 21.65/1.52/23.17

PAGE 40: 17.75/1.24/18.99 16.10/1.13/17.23
20.05/1.40/21.45 25.35/1.77/27.12

PAGE 41: 24.70/1.73/26.43 13.75/.96/14.71
12.80/.90/13.70 10.55/.74/11.29

PAGE 42: 18.95/1.33/20.28 22.35/1.56/23.91
25.15/1.76/26.91 22.50/1.58/24.08

PAGE 43: 24.35/1.70/26.05 22.70/1.59/24.29
19.60/1.37/20.97 23.35/1.63/24.98

PAGE 44: 1) .08/.36/14.06
2) .83, .87, 1.02, .68, 2.46
3) 7.92, 9.58, 6.47, .52, .94
4) 1.12, 1.53, 8.33, 6.62, 7.10
5) 8.25, 14.55, 12.41, 14.31
6) 16.83, 52.53, 141.11, 106.45

PAGE 45: 1) 4.15 2) 12.14
3) 27.80, 25.90, 7.80, 18.25
4) 94.20 5) 960.88

PAGE 46:

$ .13	$ .27	$ .81	$ .77
$ .40	$ .30	$ .22	$ .46
$ .75	$ .50	$ .36	$ .85
$ .15	$ .25	$ .26	$ .50
$ .74	$ .59	$ .12	$ .40
$ .42	$ .53	$ .50	$ .10

PAGE 47:

$ .32	$ .40	$ .02	$ .05
$ .61	$ .55	$ .52	$ .51
$ .41	$ .70	$ .24	$ .14
$ .30	$ .41	$ .53	$ .22
$ .71	$ .81	$ .17	$ .10
$ .12	$ .22	$ .19	$ .13

PAGE 48:

$ 1.10	$ 1.34	$ 4.30	$ 5.41
$ 4.42	$ 2.33	$ 3.37	$ 6.52
$ 3.70	$ 1.53	$ 7.11	$ 1.13
$ 3.26	$ 3.40	$ 2.12	$ 1.51
$ 3.42	$ 1.35	$ 7.10	$ 1.62
$ 1.34	$ 5.10	$ 1.10	$ 1.30

PAGE 49:

$ 2.54	$ 3.33	$ 7.05	$ 2.21
$ 3.03	$ 2.25	$ 7.05	$ .20
$ 4.05	$ 2.21	$ 1.00	$ 2.16
$ .16	$ 2.07	$ 1.25	$ .24
$ 6.20	$ 3.01	$ 1.05	$ .30
$ 5.15	$ 2.24	$ 4.15	$ 2.06

PAGE 50:

$ 2.25	$ 4.17	$ 3.13	$ 3.11
$ 12.21	$ 1.12	$ 10.22	$ 1.16
$ 21.11	$ 10.33	$ 1.05	$ 3.11
$ 1.12	$ 1.13	$ 2.22	$ 2.16
$ 2.11	$ 11.14	$ 4.11	$ 11.16
$ 13.22	$ 11.22	$ 3.01	$ 8.32

PAGE 51:

$ 61.56	$ 73.33	$ 44.01	$ 33.30
$ 73.23	$ 4.30	$ 33.35	$ 21.31
$ 51.31	$ 2.43	$ 20.30	$ 41.23
$ 74.41	$ 71.52	$ 22.64	$ 72.30
$ 13.32	$ 22.21	$ 4.61	$ 53.37
$ 21.52	$ 32.04	$ 13.25	$ 33.30

PAGE 52:

$ .18	$ .08	$ .58	$ .17
$ .08	$ .27	$ .23	$ .07
$ .49	$ .62	$ .48	$ .16
$ .07	$ .56	$ .58	$ .06
$ .38	$ .17	$ .68	$ .58
$ .68	$ .56	$ .07	$ .07

PAGE 53:

$ .49	$ .48	$ .33	$ .07
$ .41	$ .19	$ .12	$ .11
$ .14	$ .72	$ .27	$ .16
$ .18	$ .07	$ .17	$ .19
$ .26	$ .04	$ .48	$ .29
$ .08	$ .27	$ .79	$ .69

PAGE 54:

$ 4.59	$ 2.88	$ 1.69	$ .74
$ 1.79	$ 1.78	$ 2.39	$ 4.56
$ .79	$ 1.58	$ 1.76	$ 3.88
$ 3.54	$ 1.39	$ 4.67	$ .79
$ 3.56	$ 4.78	$ 3.65	$ 5.75
$ 2.76	$ .78	$ .37	$ .88

PAGE 55:

$ 3.79	$ .88	$ 2.96	$ 1.78
$ 4.79	$ 6.81	$ .79	$ 1.99
$ 2.69	$ .89	$ 1.86	$ .59
$ 3.86	$ 1.57	$ 3.39	$ 3.59
$ 2.39	$ .39	$ 2.71	$ .79
$ 1.78	$ 3.58	$ 1.38	$ 2.86

PAGE 56:

$ 35.89	$ 8.89	$ 8.84	$ 27.86
$ 8.86	$ 27.86	$ 26.83	$ 8.88
$ 13.64	$ 7.45	$ 8.87	$ 8.81
$ 64.88	$ 38.89	$ 58.37	$ 8.96
$ 4.83	$ 28.85	$ 56.36	$ 38.36
$ 8.89	$ 18.88	$ 47.36	$ 8.99

PAGE 57:

$ 35.97	$ 8.89	$ 28.08	$ 28.89
$ 2.56	$ 5.38	$ .86	$ 28.19
$ 9.09	$ 19.04	$ 12.67	$ 17.41
$ 8.89	$ 2.71	$ 66.04	$ 18.89
$ 35.89	$ 16.94	$ 15.02	$ 27.69
$ 8.91	$ 5.79	$ 1.84	$ 7.89

PAGE 58:

$ 1.12	$ 1.16	$ 3.41	$ 5.68
$ 3.71	$ 3.72	$ .71	$ .06
$ .01	$ 3.11	$ 4.88	$ .81
$ 2.81	$ 4.34	$ 3.42	$ 1.07
$ .88	$ 4.84	$ 3.02	$ .06
$ .54	$ 2.17	$ 5.18	$ 4.31

PAGE 59:

$ 58.18	$ 19.11	$ 79.35	$ 14.45
$ .05	$ 23.44	$ 67.75	$ 58.05
$ 37.11	$ 49.54	$ 35.88	$ 28.02
$ 17.11	$ 9.54	$ 55.88	$ 8.02
$ 18.55	$ 59.02	$ 29.16	$ 10.01
$ 28.11	$ 55.56	$ 25.82	$ 77.02

PAGE 60:

$ 4.01	$ 50.05	$ 12.18	$ 31.03
$ 17.16	$ 13.07	$ 10.01	$ 27.64
$ 64.02	$ 5.02	$ 44.65	$ 21.09
$ 54.58	$ 62.01	$ 5.05	$ 42.03
$ 6.09	$ 24.45	$ 11.02	$ 17.88
$ 1.28	$ 1.12	$ 38.64	$ 30.61

PAGE 61:

$ 6.09	$ 18.67	$ 45.18	$ 7.87
$ 38.19	$ 3.78	$ 6.83	$ 7.64
$ 48.23	$ 9.01	$ 31.84	$ 40.80
$ 55.78	$ 14.79	$ 27.83	$ 52.88
$ 37.68	$ 53.89	$ 27.66	$ 8.12
$ 20.79	$ 14.89	$ 33.72	$ 22.67

PAGE 62:

1)	$ .04	9)	$ 1.02
2)	$ .36	10)	$ .65
3)	$ .24	11)	$ 5.50
4)	$ .07	12)	$ 2.50
5)	$ .28	13)	$ 1.82
6)	$ .97	14)	$ 3.20
7)	$ .47	15)	$ 4.55
8)	$ 3.71	16)	$ 3.08

PAGE 63: 1) $ 3.80 2) $ 2.35 3) $ .35
4) $ .40 5) $ .25 6) $ 1.85
7) $ 3.45

PAGE 64: 1) $ 3.39 2) $ .60 3) $ 2.06
4) $ 6.00 5) $ .79 6) $ 15.05
7) Pecan Banana Split, $ .20

PAGE 65: 1) $ 2.00 2) $ .55 3) $ 5.75
4) $ 7.05 5) $ .35 6) $ 3.30

PAGE 66: 1) $ 7.20/$ 2.80 2) $ 7.30/$ 2.70
3) Hot Fudge Sundae, $ .15
4) $ 8.10/$ 11.90
5) $ 7.75/$2.25
6) John, $ .55

PAGE 67: 1) $ 25.90/$ 4.10 2) Larry, $ .60
3) Lisa, $ .40 4) $ 7.30/$ 2.70
5) $ 10.90 6) $ 13.25

PAGE 68: 1) 36 2) 7 3) $ 3.52 4) $ 40.75
5) $ 80.50 6) $133.80

PAGE 69: 19.80/1.39/21.19, Change 3.81
13.95/.98/14.93, Change .07
22.40/1.57/23.97, Change 1.03
13.30/.93/14.23, Change 5.77

PAGE 70: 13.75/.96/14.71, Change .29
13.25/.93/14.18, Change 5.82
15.45/1.08/16.53, Change 3.47
12.05/.84/12.89, Change 2.11

PAGE 71: 9.45/.66/10.11, Change 4.89
16.00/1.12/17.12, Change 2.88
12.25/.86/13.11, Change 1.89
17.05/1.19/18.24, Change 1.76

PAGE 72: 7.50/.53/8.03, Change 1.97
13.25/.93/14.18, Change 5.82
16.60/1.16/17.76, Change 2.24
15.70/1.10/16.80, Change 3.20

PAGE 73: 14.15/.99/15.14, Change 4.86
16.30/1.14/17.44, Change 2.56
22.95/1.61/24.56, Change 5.44
21.00/1.47/22.47, Change 17.53

PAGE 74: 22.25/1.56/23.81, Change 1.19
24.05/1.68/25.73, Change 4.27
14.85/1.04/15.89, Change 4.11
10.40/.73/11.13, Change 3.87

PAGE 75: $ 4.60

$ .84	$ .99	$ 4.47	$ 9.59
$ 1.20	$ 8.42	$ 9.21	$ 12.14
$ 14.33	$ 54.53	$101.67	$ .32
$ 7.89	$ 2.12	$ 40.05	$ 42.48
$ 29.09	$ 67.05	$ 53.89	$ 3.34

PAGE 76: 1) $ 9.60/$ 11.25 2) $1531.26
3) $ 7.20 4) $ 3.45
5) $ 8.00, $ 12.00 6) $113.69
7) $ 5.25